A DAY IN AN ECOSYSTEM

24 HOURS IN THE DESERT

VIRGINIA SCHOMP

New York

Published in 2014 by Cavendish Square Publishing, LLC
303 Park Avenue South, Suite 1247, New York, NY 10010

First Edition

Website: cavendishsq.com

CPSIA Compliance Information: Batch #WS13CSQ

Library of Congress Cataloging-in-Publication Data

Schomp, Virginia.
24 hours in the desert / Virginia Schomp.
p. cm. — (a day in an ecosystem)
Includes bibliographical references and index.
Summary: “Take a look at what takes place within a 24-hour period in a desert. Learn firsthand about the features, plant life, and animals of the habitat”—Provided by publisher.
ISBN 978-1-60870-893-2 (hardcover) ISBN 978-1-62712-066-1 (paperback)
ISBN 978-1-60870-900-7 (ebook)
1. Desert ecology—Juvenile literature. I. Title. II. Title: Twenty four hours in the desert.
QH541.5.D4S36 2013
577.54—dc23
2011041777

Editor: Peter Mavrikis
Art Director: Anahid Hamparian
Series Designer: Kay Petronio
Photo research by Alison Morretta

Printed in the United States of America

CONTENTS

DAWN 5

MORNING 9

AFTERNOON 19

EVENING 27

NIGHT 35

FAST FACTS 43

GLOSSARY 44

FIND OUT MORE 46

INDEX 47

DAWN

THE SUN creeps out from behind a golden sand dune. The air is mild. The ground beneath your feet is soft and dry. A vast sea of sand stretches out as far as you can see. It is time to start your day's adventure in the Sahara Desert.

What do you think of when you hear the word **desert**? Chances are, you picture a place that looks like this part of the Sahara. You might see a land that is hot and dry and covered with sand. You might imagine walking for miles without meeting another living thing.

There is some truth to that picture. All deserts *are* dry. Some scientists define a desert as an area that gets less than 10 inches (25 centimeters) of rain a year. That makes deserts the driest places on earth.

But aside from their dryness, deserts are not all alike. Some are blazing hot, and some are freezing cold. Some have miles and miles of sand piled up in sand dunes. Others have more rocks than sand. The land may

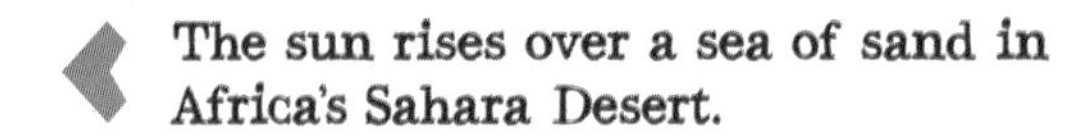

The sun rises over a sea of sand in Africa's Sahara Desert.

AND THE WINNER IS . . .

The Sahara takes the prize for World's Largest Desert. It covers 3.5 million square miles (9 million square kilometers) in northern Africa. That is almost as big as the whole United States.

be flat, or it may be carved into mountains and valleys. There may be strange rock formations that look like giant fingers or tall arches or the ruins of ancient temples.

A scorpion scurries across the hot desert sand with its poisonous stinger at the ready.

There is another surprising thing about deserts. They may look "deserted," but they are actually teeming with life.

Listen! Do you hear something rustling in that patch of dry grass? It could be a deadly horned viper. Or a fat sand rat. Or maybe a little fennec fox, settling down to rest after a night of hunting. An amazing variety of plants and animals make their home here. Let's take a walk and find out how they survive the harsh conditions of the Sahara Desert.

A COLD DESERT?

Antarctica measures 5.5 million square miles (14 million square kilometers). This cold continent gets mostly snow instead of rain. The snowfall equals less than 8 inches (20 centimeters) of rain a year. Some scientists say that makes Antarctica a desert. Others disagree. They say Antarctica is so different from other deserts that it belongs in a group by itself.

MORNING

THE FIRST THING you notice in the desert is the wind. It never seems to stop blowing. It fills the air with millions of tiny grains of dust and sand. As you cross the desert, you will want to keep your nose and mouth covered. And don't forget your sunglasses!

It is the wind that makes all those sand dunes. Dunes come in many different shapes, depending on the force and direction of the wind. They may look like small ripples or towering ocean waves. They may even look like stars.

Climb to the top of a dune. What do you see? Do the sands seem to stretch on forever? In fact, only about one-fifth of the Sahara Desert is covered with sand. These sand seas are called **ergs**. *Erg* means "dune field" in Arabic, the main language of North Africa.

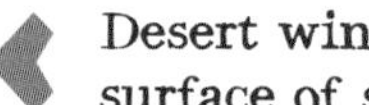

Desert winds can form ripples on the surface of sand dunes.

RUNAWAY DUNES

Did you know that sand dunes can move? The grains of sand blow up one side of the dune and roll down the other. Slowly the dune creeps forward. Over time, the drifting sands can bury roads, fields, and villages. People sometimes try to hold back the wandering dunes by putting up "sand fences."

The rest of the Sahara is a mixture of landscapes. There are broad, flat plains covered with gravel. There are large stone **plateaus**. The steep Atlas Mountains mark the northern edge of the desert. Deep within the Sahara are other mountain ranges. Sometimes the tallest mountains are capped with snow. That is a strange sight in this sun-baked land!

It is still early morning, but it is already getting hot. Look around and you will see why. There are no clouds to block the sun's rays. There are no cool fields of grass or shady trees. All you can see are sand and rocks, which heat up quickly in the blazing sun. On a summer day, temperatures can climb over 130 degrees Fahrenheit (55 degrees Celsius). That is hot enough to burn the soles of your feet. So keep your shoes on!

The heat will make you sweat. The more you sweat, the more water your body loses. To keep safe, you must drink plenty of water. Without water or shelter, a person can survive only a few hours in the extreme heat of the desert.

You will have to carry your water with you. The Sahara is one of the driest places on the planet. Most of this desert gets less than 4 inches (10 centimeters) of rain a year. In the super-dry center of the Sahara, it may not rain for years at a time. When it does rain, much of the water **evaporates** in the heat. Sometimes it is so hot that the raindrops dry up before they even hit the ground!

As you hike across the dunes, you pass some patches of spiky-looking grass. A small shrub sticks its head out of the ground. You know that all plants need water to live. So how can anything grow in the Sahara?

The answer is **adaptation**. Over thousands of years, plants have adapted, or developed ways to survive in the dry desert. The roots of that spiky grass have a thick coating to keep

HOT AND HOTTER

The Sahara Desert holds the record for the highest air temperature ever measured—136°F (58°C). The second-highest temperature was 134°F (57°C). It was recorded at a ranch in Death Valley, California, on July 10, 1913. A man at the ranch said, "It was so hot that swallows in full flight fell to the earth dead."

MADE IN AMERICA

The most famous plant with spines is the cactus. But you will not find this prickly plant in the Sahara Desert. Except for a very few types, the cactus is American. It grows mainly in the deserts of North, Central, and South America.

Succulent shrubs have adaptations that help them survive in the hot, dry desert.

water from leaking out. That small shrub is a succulent—a type of plant that stores water in its roots, stems, or leaves. Many succulents have roots that spread out just beneath the surface. These shallow roots quickly soak up the morning dew or a light rain shower. Other desert plants have deep roots to seek out water hidden far below ground.

Look closely at the shrub's leaves. They are small and rounded. Other desert plants may have stiff, narrow leaves called "spines." Water can evaporate quickly from leaves that are wide and flat. The leaves of desert plants are designed to hold on to every drop.

Plants have other ways of adapting to desert conditions. They usually grow far apart. That way, they do not have to compete for water. Most do not grow very tall. Small plants need less water to survive. There is not much food in the desert for plant-eating animals, so plants have ways to make sure they do not become some critter's lunch. They may smell bad or taste bad. They may be poisonous. They may have tough bark or nasty thorns.

Some plants do not even try to deal with day-to-day life in the desert. Instead, they hide. During the long dry periods, these plants survive underground as seeds.

When the rain finally comes, the seeds burst to life. Plants shoot up so fast you can almost watch them grow. They cover the desert with flowers. The flowers scatter new seeds on the ground. A few weeks later,

PLAYING DEAD

Mosses are some of the smallest desert plants. These tiny plants grow mainly on rocks. How do they survive in the desert? They dry out completely between rainstorms. In one museum, scientists watered a desert moss that had gone 250 years without water—and it started to grow!

Bright yellow flowers carpet the desert after a rainstorm.

the plants die. The seeds settle down into the soil, and the cycle starts all over again.

The rain also brings a burst of animal life. Some desert creatures begin their lives only after a heavy rain. Insect eggs hatch in the muddy pools left by the storm. Soon swarms of winged termites fill the air. Moths and butterflies flit from plant to plant, feeding on the sweet **nectar** made by the flowers.

A female spadefoot toad keeps an eye on all the busy insects. She has been dozing underground for months. When the rains came, she dug her way back to the surface. She mated and laid her eggs. The eggs hatched, filling the pool with wriggling tadpoles. The tadpoles will take only a few weeks to grow up. By the time the ground dries out, they are already adult toads. They burrow into the soil to wait for the next rainstorm.

Other desert animals are active year-round. But even these creatures are especially lively after the rain. Birds and snakes feast on all the new insects. Sand rats and gerbils scurry around, gathering seeds. Tortoises, hares, and gazelles munch on the plants and flowers. Jackals and foxes hunt for smaller animals. All these creatures are stocking up on energy for the long dry spell ahead. The stored seeds and extra body fat will help them survive when food is hard to find.

You would have to look hard to find any of those lively animals right now. The sun is climbing, and few creatures are out and about. You might see a golden eagle soaring overhead. Its shadow passes over a tiny lizard called a

A sandfish lizard

SPEEDY SAND SWIMMERS

The sandfish gets its name because it wiggles through sand like a fish swimming through water. As it "swims," it listens for insects. When a meal wanders by, it zips to the surface. This little lizard can move at 6 inches (15 centimeters) a second. That speed helps it catch its **prey**—and escape from snakes and other **predators**.

sandfish. The sandfish is hunting for lunch. It dives down and moves just under the surface of the sand. Snap! The lizard shoots back up and nabs an unlucky ant.

Just like the plants, all these creatures are well adapted for desert life. The eagle has a naturally high body temperature. That helps it stay comfortable in the heat. Ants and other insects have a waxy outer coat that helps keep the water in their bodies. Thick, watertight skin keeps lizards, snakes, and other **reptiles** from drying out.

Reptiles are cold-blooded animals. That means their body temperature changes with the temperature around them. In the early morning and late afternoon, a reptile warms up in the sun. During the hottest part of the day, it keeps cool in the sand or shade.

Other desert animals avoid the sun completely. They are **nocturnal**. Nocturnal animals are active at night and keep still during the day. Some rest in

Cold-blooded reptiles like this lizard must warm their bodies by basking in the sun.

the shade of rocks or plants. Some hide underground in cool burrows. At night, these animals will begin to stir. They will come out to look for food. And if they are not careful, they may become a meal themselves.

AFTERNOON

IT IS MIDDAY. The sun is burning right over your head. The desert is quiet. Nearly all the animals have taken shelter from the heat.

You might want to follow their lead. Just be careful where you sit. A horned viper may lie in wait under the sand. This snake buries itself so that only its eyes and horns stick out. When a careless mouse or lizard wanders by, it strikes. Its sharp fangs inject deadly **venom** into its prey.

Watch out for scorpions, too. These small **invertebrates** may hide under rocks during the day. The scorpion has a stinger on the end of its tail. It uses the stinger to shoot venom into its prey. The scorpion feeds mostly on insects, spiders, and other scorpions. Its sting will not usually kill a person. It is very painful, though. So you definitely want to keep out of this clawed creature's way!

A deep-rooted acacia tree bakes in the noontime sun.

The sand-colored horned viper is one of the most venomous desert snakes.

If you are lucky, you may find shade in a **wadi**. Wadis are steep-sided riverbeds in the desert. After heavy rains, they fill up with fast-flowing water. Most of the time, though, the wadi is dry. A few **acacia** trees may survive between rainstorms. Acacias have deep roots and fat trunks that hold on to water.

If you are *very* lucky, you might find an **oasis**. An oasis is a cool, green place where plants grow all year round. What causes an oasis? Even in the desert, there is often water underground. The water may have fallen as rain hundreds of miles away. It flows beneath the desert, trapped between layers of rock. When it comes to the surface, an oasis forms. Sometimes this happens naturally, where the ground dips down. Sometimes people dig wells to tap into the hidden water supply.

When you first spot the oasis, it looks like a speck of green in an ocean of sand. Walk closer. Can you see the date palms? These tall trees wave gently in the desert breeze. Sit down in the shade! Enjoy a drink from the well! You might even

Date palms line the shores of a large blue lake in a desert oasis.

AN ANCIENT DRINK

If you go to Australia, do not be surprised if someone offers you a really old drink. People in the Simpson Desert get their water from a huge underground pool called the Great Artesian Basin. Some of the water in the basin came from rain that fell 2 million years ago.

climb up to pick one of the small brown fruits hanging from the crown of the palms. The dates are sweet and tasty—but watch out for the tree's sharp thorns!

Your oasis is tiny. It has just a deep well and a few palm trees. There are also about ninety large oases in the Sahara, where people have built villages and towns. These people have learned to get the most out of the land and water. They make their homes from mud bricks or stone. They

Palm trees flourish in the narrow strip of fertile soil between the desert sands and the Nile River.

build small dams to collect the groundwater and rainwater. Earth and stone canals channel the water to their crops.

The most important crop in the oasis is the date palm. Smaller trees may include orange, lemon, peach, and fig. The trees shelter fields of onions, carrots, maize (corn), wheat, barley, **millet**, and other crops.

Crops also grow beside the rivers. Two major rivers cross the Sahara. The Nile River is the longest river in the world. It flows for more than 4,100 miles (6,600 kilometers) through northeast Africa. The Niger River cuts through the southwest corner of the desert.

These two rivers provide water for drinking, cooking, washing, and farming. People fish in the rivers. Boats carry goods and passengers. Dams harness the power of the flowing waters. The Aswan High Dam on the Nile River is one of the world's largest dams. It produces electricity for millions of homes and businesses in Egypt.

A desert mirage

A TRICK OF THE LIGHT

When is a lake not a lake? When it is just a mirage.

A mirage happens when the air near the ground is much hotter than the air at eye level. As light waves enter the hot air, they bend upward. You see a reflection of the blue sky—but to your confused brain, it looks like a pool of water.

A Sahara sandstorm

LONG-DISTANCE DUST

Sometimes the desert wind picks up speed. A fierce dust storm or sandstorm develops. The clouds of swirling dust or sand make it hard to breathe and impossible to see. The dust can ride the winds all the way across the Atlantic Ocean. Dust from the Sahara Desert may even end up on a beach in Florida!

Not all desert people settle at an oasis. There are also many people who live as **nomads**. Nomads move from place to place, in search of food and water for their animals.

The most famous nomads of the Sahara Desert are the **Tuareg**. You might meet a group of Tuareg men at the oasis. You will recognize them by their clothes. The nomads wear deep-blue robes and turbans. They cover their faces with a veil. Their clothing protects them from the sun, wind, and swirling sand.

The Tuareg have come to the oasis for water. A one-humped camel pulls the rope that raises the bucket from the well. The camel is highly adapted for desert life. It can go two weeks or longer without a drink. People used to think it stored water in its hump. Now we know that is not true. The reason camels can go so long without drinking is that their bodies are very good at holding on to the water they take in. They are also champion drinkers. A thirsty camel can slurp up 25 gallons (95 liters) of water in ten minutes. That is ten times more than a person can drink!

The camel has many other adaptations. Its thick coat protects it from the sun and heat. Its wide, leathery feet help it walk in the soft sand. When dust and sand blow, the camel can close its nostrils. Long eyelashes shield its eyes. If a grain of sand does get in, the camel has an inner eyelid that acts like a windshield wiper to clean it out. This lid is very thin. In a blinding sandstorm, a camel can see with its eyes shut!

The rest of the Tuareg's herd includes sheep and goats. All the animals in the herd provide the nomads with meat and milk. The wool, fur, and hides are used to make tents, rugs, blankets, clothes, shoes, and other supplies. Even the droppings are useful. No wood? No worries! Dried camel dung makes a good fire!

EVENING

THE HEAT of the day is beginning to fade. It is time to move out. Why not travel for a while with the Tuareg?

The nomads are constantly on the move, so everything they own must be easy to carry. Easy for a camel, that is! When the Tuareg move camp, they load their camels with all the tents, carpets, blankets, bowls, bags, cooking pots, food, and water. A camel can carry 500 pounds (230 kilograms) for hundreds of miles across the scorching sands. No wonder this sturdy animal is known as the "ship of the desert"!

Will you climb into the saddle? Riding a camel is wobbly, but it is a great way to see the sights. Up above the hot sands, there is a constant breeze. The low, rolling dunes stretch out before you. The camels plod along at a slow, steady pace. There is no rush hour here. Moving quickly in the desert would just waste water and energy.

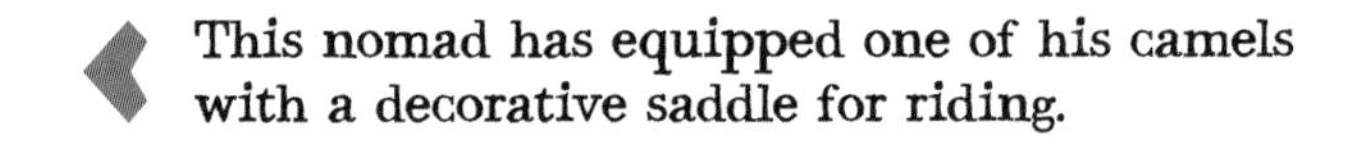

This nomad has equipped one of his camels with a decorative saddle for riding.

The nomads do not wander around aimlessly. They know exactly where they are going. They know where the grasses grow on the borders of the desert, even during the hottest part of the summer. They know where the rains will bring a brief burst of life to the dry land.

The paths to these places are not marked on a map. How do the herdsmen find their way? By night, they follow the stars. During the day, they "read" the desert. A nomad can plot his course by a lone tree, a large rock, or changes in the soil and shadows.

Women prepare a meal at a Tuareg camp near the southern edge of the Sahara.

When the nomads reach fresh grazing lands, they set up their tents. They will stay in the area for just a few days. This keeps the animals from overusing the land. When the herd moves on, the plants will recover. After the next rains, the pastures will turn green again.

The Tuareg live simply. Nearly everything they need comes from their animals. But there are some things the herds cannot provide. Sometimes the nomads must trade with the people in towns and villages. These settlements may be at a large oasis or in a strip of grasslands south of the Sahara called the **Sahel**.

Your nomad guides have come from an oasis town. They brought their milk, meat, animals, and handmade cloth and blankets. They traded their goods for grain, sugar, and tea. Now they are returning to the camp where their families are waiting.

CAMEL TRAINS

For hundreds of years, Tuareg guides led **caravans** across the Sahara. There were thousands of camels in those long desert "trains." They carried goods such as salt, gold, ivory, and ostrich feathers. These goods were traded with cities to the north and south. Today trucks and planes carry most trade goods across the desert.

A town in the Sahel is a bright, busy, colorful place on market day.

The trip takes you over a constantly changing landscape. You leave the sands to cross over a flat area covered with coarse pebbles. This is the ***reg***, or gravel plains that cover nearly three-quarters of the Sahara. Each time you look, the colors of the stones are different. They are gray, black, or white, with patches of green and blue.

After a while, you pass some large boulders. They have been carved into amazing shapes by the wind. Off in the distance, a jagged plateau reaches for the sky. The plateau areas of the Sahara are called ***hammada***. In the

Camels cross the bare, stony part of the desert known as the *hammada*.

hammada, the wind has scoured away all the dirt, sand, and pebbles. All that is left is bare rock.

There are few signs of life in these rocky parts of the desert. A patch of grass clings to a boulder. A few small bushes straggle along the bottom of a dry riverbed. A flock of sandgrouse flies noisily overhead. These desert birds can fly many miles in search of water. The male sandgrouse has special feathers that soak up water like a sponge. After the bird drinks, it flies back to its nest and brings water to its thirsty chicks.

The sun is beginning to sink. A colorful butterfly flits across your path. Soon the rocky plain gives way to a line of sand dunes. You come to a lonely-looking acacia tree. Here the Tuareg stop for their evening meal. They gather dry twigs and rub sticks together to make a fire.

Dinner is a thick white soup and flatbread. The soup is made by adding water to a powder of dried millet, onions, and dates. It is cooked in a pot over the fire. While it bubbles, the cook kneads water into millet flour. He forms the dough into a flat disk. He places the loaf in the hot sand by the fire. He covers it with glowing embers. After a few minutes, he turns it over to cook the other side. Then he digs it up and dusts off the ashes and sand. The men pull the bread apart and shred it into the soup. After the

A nomad bakes bread in the hot sand near a campfire.

meal, you sip a small glass of tea. Sweet mint tea is a favorite drink all over the Sahara.

The sun sets in a blaze of color. The nomads get ready to move on. The cool of the evening is a good time to travel in the desert. It is also the perfect time to see some interesting desert dwellers. As the temperatures drop, these creatures will come out of their hiding places and bring the desert to life.

THE DESERT IS COOL!

After the sun sets, the desert cools down quickly. There are no clouds to keep the heat from escaping into the sky. Temperatures in the Sahara can drop 50°F (10°C) or more. All of a sudden, the world's biggest oven is a giant refrigerator!

NIGHT

THE MOON has risen. It lights your way across the cool sands. A desert eagle owl hoots in the night, disturbed by the steady plodding of your camels.

It is feeding time for the desert's nocturnal animals. The owl uses its sharp hearing to zero in on its prey. Suddenly, it swoops down to the sand. Its sharp claws grab a scorpion by the tail. The owl nips off the deadly stinger before flying off with its crunchy meal.

Other desert hunters are also on the prowl. You shine your flashlight and surprise a fennec fox. Like the owl, this tiny fox hunts mainly by hearing. Its big ears twitch as it listens for the movements of its prey. The fennec fox dines on small **rodents** and insects. It gets most of the water it needs from the bodies of the creatures it eats.

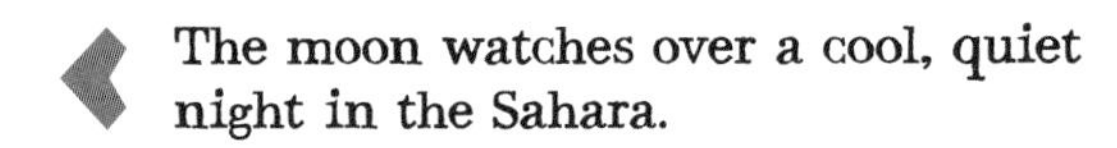

The moon watches over a cool, quiet night in the Sahara.

WHAT BIG EARS YOU HAVE!

Oversized ears help the fennec fox hunt its prey. Those big ears have another use, too. Heated blood flows through a web of blood vessels in the ears. The blood is cooled by the breeze before it returns to the body. That keeps the fox from getting too hot.

Quick! There goes one of the fox's favorite snacks—a jerboa. This little rodent has ventured out from its burrow to look for water-rich plants and seeds. The fox leaps, but the jerboa is faster. With its long hind legs, it can hop like a kangaroo. It bounds off to find a safe place to hide. The hungry fox will have to look somewhere else for a meal.

Heart-shaped tracks in the sand show that a larger animal has been this way. Soon you see a small herd of dorcas gazelles. These small sand-colored gazelles have passed the day in the shade. Now they are munching on a patch of desert herbs. They will spend the night searching for other small plants, twigs, and leaves. They choose their food carefully. They are looking for plants that are full of water. A dorcas gazelle can go its whole life without ever taking a drink.

At last, your small caravan reaches the Tuareg camp. You see a small group of tents made of goatskins stretched over wooden poles. Several families live together in the camp. The families are all related to each other. Each family—mother, father, and children—has its own tent.

The women and children come out to greet you. While the men take care of the camels, one woman invites you inside her tent. You sit on a leather cushion. You drink a wooden bowl of milk. Camel or goat's milk is the nomads' traditional gift to a stranger.

You learn that a Tuareg woman owns her family's tent. When the nomads move, she is in charge of setting it up in the new camp. Women are also responsible for cooking the family's food. They make the clothing, rope, and other things the family needs. Tuareg women are often better educated than the men. They are the ones who teach the children and pass down the stories of their people.

All but the youngest children have chores. Girls help their mothers with the cooking, sewing, and other tasks. Boys take care of the smaller animals. When they get older, they help their fathers with the camels.

Children have another important job. They fetch the family's drinking water. The Tuareg camp is set up a couple of miles from a well. Each day, the children load the empty water skins on the family's donkey. They lead the donkey to the well. They watch while herdsmen give water to

Tuareg herdsmen haul a bucket filled with water from a desert well.

the noisy flocks of goats and groups of camels. Then they haul up the water and pour it into the water skins. It is hard work but fun. The well is a meeting place where young people can talk, laugh, and play while they wait for their turn at the water.

The night has turned bitter cold. You sit by a crackling fire, under a sky bright with stars. There you meet another visitor to the Tuareg camp. It is a scientist from an African conservation group. He is studying the wildlife of the Sahara Desert.

The scientist tells you that the desert **ecosystem** is always changing. For millions of years, these changes happened very slowly. But today, people are doing things that change the land very quickly. They pump lots of water to build cities in the desert. They graze large herds of cattle in the Sahel. They drill for oil. They mine for minerals. They build hotels for tourists.

These activities have brought riches to some desert nations. But they also bring dangers. Too much pumping can use up the groundwater. Overgrazing can kill the plants and leave the soil bare. Mining and oil drilling cause pollution. Tourists may damage the beautiful lands they have come to admire.

All this can lead to something that scientists call **desertification**. Desertification happens when human activities turn productive land into desert. An oasis dries up, and all the plants die. Fertile soil in the Sahel turns to dust.

People's actions can also threaten desert animals. The scientist has been tracking a mother cheetah and her cubs. Cheetahs once lived all over Africa. Today there are only a few thousand left. People have taken

When an oasis dries up, the trees die and busy desert towns fall into ruins.

over many of the open spaces that are the cheetah's natural home. They have killed the antelopes that the big cat hunts for food. They have hunted the cheetah for its fur.

Scientists study cheetahs and other endangered animals to find ways to protect them. They are also studying the changing deserts. They are working on better ways for people to manage these lands. By using the land and water wisely, people can save the deserts. They can protect the plants, animals, and people that live in the world's driest places.

You have talked through the night. The darkness is fading. Can you see all the tracks in the sand? The foxes, rodents, and other nocturnal animals have been busy. Now they are returning to their shady hideaways.

As the animals go to bed, the nomad camp begins to stir. Children gather sticks for the fire. Their mothers mix millet with dates and milk to make porridge. The men will eat first. Then they will set out to tend the herds of camel.

Outside the camp, daytime creatures are stirring, too. Insects hop and crawl and fly. A thorny lizard crawls out onto a rock. It basks in the rising sun, lashing its prickly tail to scare away predators. A sharp-eyed hawk perches in a tree, ready to swoop down on its breakfast. Your time in the Sahara Desert has come to an end. But out on the oceans of sand, a new day's adventure is just beginning.

An ancient rock painting of giraffes

THE ONCE-GREEN SAHARA

The Sahara was not always a desert. How do we know? Scientists exploring the sands have found the **fossils** of animals that belong in lakes, woods, and grasslands.

The rocks also hold clues. Thousands of years ago, people painted hunting scenes on rock walls. Their paintings show that elephants, giraffes, and crocodiles once roamed a green Sahara.

WHERE ARE THE DESERTS?

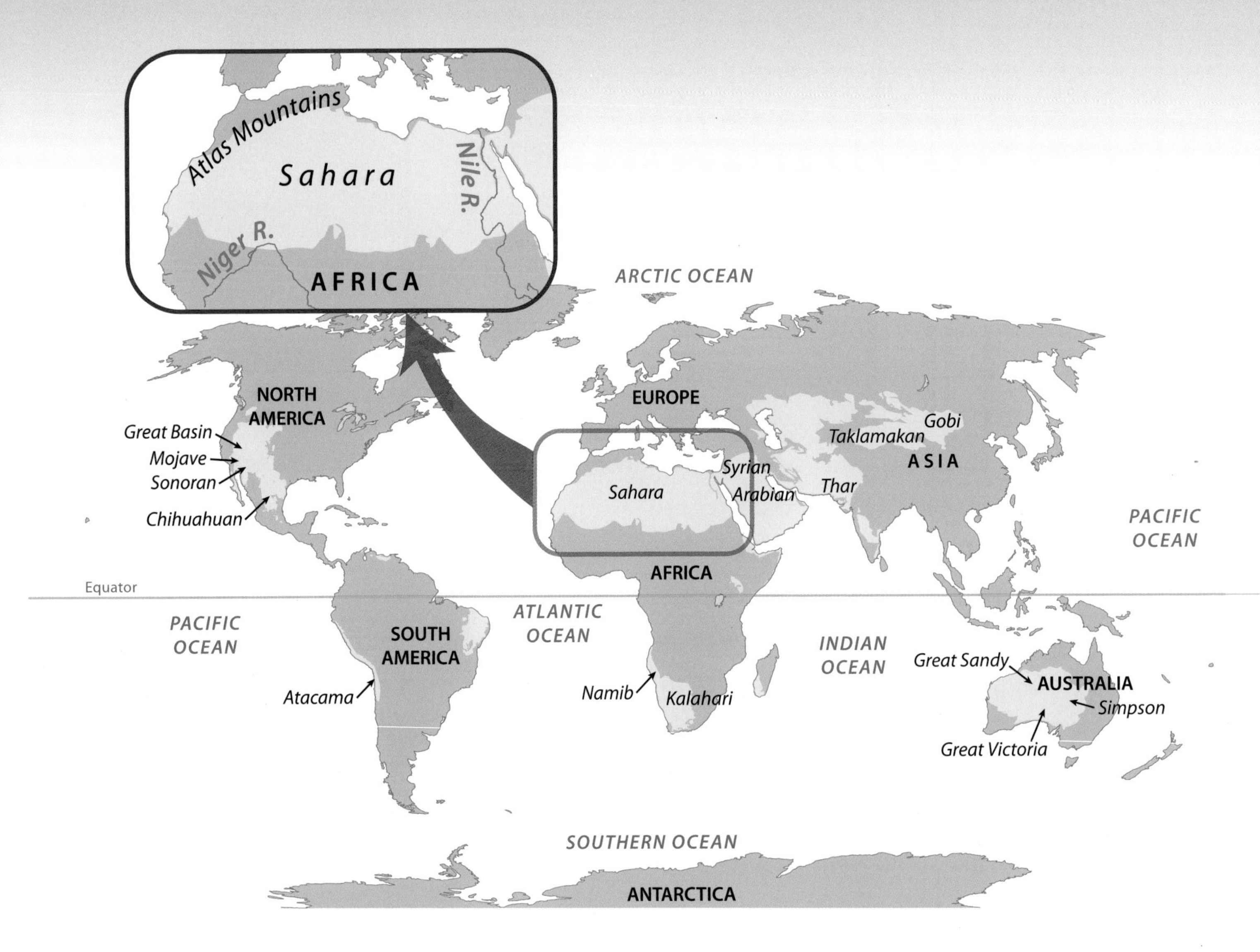

Deserts cover about one-fifth of the world. This map shows the location of some of the largest deserts.

FAST FACTS ABOUT THE SAHARA DESERT

SIZE: 3.5 million square miles (9 million square kilometers).

LOCATION: The Sahara runs through eleven countries or territories in northern Africa: Algeria, Chad, Egypt, Libya, Mali, Mauritania, Morocco, Niger, Sudan, Tunisia, and Western Sahara.

LAND SURFACES: About 70 percent is covered with *regs* (gravel plains), about 20 percent with *ergs* (sand seas). The rest is mainly mountains and *hammadas* (stone plateaus).

TEMPERATURES: Average annual temperature is 86°F (30°C). Highest temperature ever recorded was 136°F (58°C) at El Azizia, Libya.

RAINFALL: Three-quarters of the Sahara gets less than 4 inches (10 centimeters) of rain a year.

MAJOR WATER SOURCES: The Nile and Niger rivers.

PLANTS: About 500 different species of plants, including trees, shrubs, herbs, grasses, and mosses.

ANIMALS: *Mammals* include dorcas gazelles, Barbary sheep, jackals, cheetahs, fennec foxes, North African hedgehogs, African porcupines, Cape hares, sand rats, mice, gerbils, and jerboas. *Birds* include falcons, bustards, hawks, golden eagles, sandgrouse, wheatears, and desert eagle owls. *Reptiles* include horned vipers, land tortoises, geckos, chameleons, and sandfish and other lizards. *Amphibians* include salamanders and spadefoot toads. *Invertebrates* include scorpions, tarantulas, camel spiders, and insects such as ants, termites, beetles, locusts, moths, and butterflies.

POPULATION: About 4 million people. More than half are nomads. The rest live mainly in oases, which are fertile areas near water.

GLOSSARY

acacia (uh-KAY-shuh)—A kind of tree or shrub that is well adapted for life in the desert.

adaptation—The process of adapting, or changing to survive under the conditions in a certain environment.

caravan—A large number of people and pack animals traveling together across a great distance. In modern times, trucks may replace the animals.

desert—A large area of land that is so dry it is hard for most plants and animals to survive there. Deserts are sometimes defined as places that get less than 10 inches (25 centimeters) of precipitation (rain or snow) a year.

desertification—The creation of desertlike conditions in once-fertile lands, through human activities such as overuse of the water.

ecosystem—An area that is home to a particular group of plants and animals, which are specially suited to living in that environment. An ecosystem includes all the living things of the area plus all the nonliving things, such as the water, soil, and rocks.

ergs—Large desert areas covered with sand.

evaporates—Dries up. Water evaporates when it is changed into vapor by the sun.

fossils—The hardened remains of long-dead animals and plants.

hammada—A desert area made up mainly of bare, rocky plateaus.

invertebrates—Animals without a backbone, such as insects and scorpions.

millet—A kind of grass grown for its grain.

nectar—A sweet liquid made by flowers.

nocturnal—Active mainly at night.

nomads—People who move from place to place, looking for food and water for their animals.

oasis (oh-AY-sus)—A green, fertile area in the desert. The plural is *oases* (oh-AY-seez).

plateau—A large area of flat land that is raised above the surrounding landscape.

predators—Animals that hunt and kill other animals for food.

prey—An animal that is hunted by a predator.

***reg* (rej)**—A broad desert plain covered with gravel.

reptiles—Animals that have scaly skin and, in most cases, lay eggs.

rodents—Small animals that have big front teeth for gnawing, such as mice and rats.

Sahel—A narrow band of grasslands on the southern border of the Sahara Desert.

Tuareg (TWAH-reg)—A nomadic people of the Sahara.

venom—The poison of some snakes, scorpions, and other animals.

wadi (WAH-dee)—A channel carved out by a stream or river in the desert. Wadis are usually dry.

FIND OUT MORE

Books

Barber, Nicola. *Living in the Sahara.* Chicago: Raintree, 2008.

Ceceri, Kathy. *Discover the Desert: The Driest Place on Earth.* White River Junction, VT: Nomad Press, 2009.

Green, Jen. *Life in the Desert.* New York: Gareth Stevens, 2010.

Hyde, Natalie. *Desert Extremes.* New York: Crabtree, 2009.

Wojahn, Rebecca Hogue, and Donald Wojahn. *A Desert Food Chain.* Minneapolis: Lerner Publications, 2009.

Websites

Biomes of the World: Desert

http://mbgnet.net/sets/desert/index.htm

Click on the "Desert Topics" links to learn about deserts and the plants and animals that live there. This colorful website is presented by the Missouri Botanical Garden.

Camel Pictures and Facts

http://fohn.net/camel-pictures-facts/index.html

Here you will find lots of information plus some great photographs of the "ship of the desert."

Enchanted Learning: Desert Habitats

http://enchantedlearning.com/biomes/desert/desert.shtml

This Enchanted Learning site has information, pictures, and activity sheets for learning about deserts and desert animals.

World Biomes: Desert

http://kids.nceas.ucsb.edu/biomes/desert.html

This Kids Do Biology site explores the plants, animals, and people of the desert.

INDEX

Page numbers in **boldface** are illustrations.

acacia tree, **18,** 20
adaptation, 11–13
animal life, 14–17, **16, 17,** 19–20, **20,**
24–25, 32, 39–41, 43
Antarctica, 7
Aswan High Dam, 23

cactus, 12
camels, 24–25, **26,** 27, 29, **31**
caravans, 29
cheetahs, 39–40
climate, 10–11, 33, 43
crops, 23

date palms, **21,** 21–22, 23
Death Valley, 11
desertfication, 39
dunes, **8,** 9, 10

ecological concerns, 39–40
ecosystem, 39
ergs, 9

fox, fennec, 35–36, **36**

gazelles, dorcas, 36
Great Artesian Basin, 21

hammada, **31,** 31–32
horned viper, 19, **20**

jerboa, 36

landscapes, 10

map, **42**
mirage, 23, **23**
moss, 14

Niger River, 23
Nile River, **22,** 23
nocturnal animals, 35–36, **36**
nomads, 24, 25, **26,** 27–29, **28,** 32–33,
33, 37–38, **38,** 43

oasis, 20–22, **21,** 39, **40**
owl, desert eagle, 35

people, desert, 22–24
plant life, 11–14, **12, 14, 21,** 21–22, 23, 43
population, 43

rainfall, 43
reg, 30
reptiles, 16, **16, 17**
rock painting, **41**

Sahel, 29, **30**
sand dunes, **8,** 9, 10
sandfish, 16, **16**
sandgrouse, 32
sandstorm, 24, **24**
scorpions, **6,** 19
Simpson Desert, 21

temperatures, 10–11, 33, 43
Tuareg people, 24, 27–29, **28,** 32–33,
33, 37–38, **38**

wadi, 20
water wells, 37–38, **38**
wind, 9

ABOUT THE AUTHOR

VIRGINIA SCHOMP has written more than eighty books for young readers on topics including dinosaurs, dolphins, world history, American history, myths, and legends. She lives among the tall pines of New York's Catskill Mountain region. When she is not writing books, she enjoys hiking, gardening, baking (and eating!) cookies, watching old movies and new anime, and, of course, reading, reading, and reading.

PHOTO CREDITS

The photographs in this book are used by permission and through the courtesy of:

Front cover: imagebroker.net/Superstock.

Alamy: Guenter Fischer/imagebroker, 1; Frans Lemmens/Corbis Cusp, 6; mediacolor's, 12; Bob Gibbons/Science Photo Library, 14; blickwinkel/Gemperle, 17; GFC Collection, 21; Michael Ventura, 22; Eitan Simanor, 30; Bernd Mellmann, 31; Mark Eveleigh, 33; TNT Magazine, 34; Amar and Isabelle Guillen/Guillen Photography, 36; Images & Stories, 38; Victor Paul Borg, 40. *Superstock:* Science Faction, 4, 18, 23; Robert Harding Picture Library, 8; age fotostock, 16, 41; imagebroker.net, 20; Frans Lemmens, 24, 26, 28.